William Spencer

The Effect Produced upon Respiration by Faradic Excitation of the Cerebrum in the Monkey, Dog, Cat, and Rabbit

William Spencer

The Effect Produced upon Respiration by Faradic Excitation of the Cerebrum in the Monkey, Dog, Cat, and Rabbit

ISBN/EAN: 9783337197148

Printed in Europe, USA, Canada, Australia, Japan

Cover: Foto ©berggeist007 / pixelio.de

More available books at **www.hansebooks.com**

XIV. *The Effect produced upon Respiration by Faradic Excitation of the Cerebrum in the Monkey, Dog, Cat, and Rabbit.*

By W. G. SPENCER, *M.S., M.B., Assistant Surgeon to the Westminster Hospital.*

Communicated by Professor VICTOR HORSLEY, *F.R.S.*

Received December 15, 1893,—Read January 25, 1894.

[PLATES 57–59.]

CONTENTS.

INTRODUCTION.

The object of the following investigation was to elucidate the character of the representation of respiration in the highest nerve centres. For this purpose the excitation method was employed, so that the results of the research are embodied in the effects produced on the movements of respiration when various regions of the cerebral hemisphere are excited.

By a careful regulation of the anæsthetic state in the species of animal used, by a recognition of the exact spot excited, and by employing a suitable stimulus, a constant effect upon respiration can be obtained from the cerebral cortex according to the point stimulated. And the same result can also be obtained along a line joining the area discovered in the cortex with the medulla oblongata.

It is absolutely necessary, in order to get constant results, that the degree of anæsthesia employed should be accurately recognized. This has apparently not been done before. While some previous observers have, in fact, not used an anæsthetic at all, others have employed considerable doses of morphine or chloral. Experiments by different observers on non-anæsthetized animals have yielded variable results, due to the fact that other disturbing factors have not been taken into account. One of these is apnœa; for instance, any excitation of a sensitive part may cause acceleration of the respiratory rhythm, and, as a consequence of this acceleration, slowing, or even a short arrest. During the prevalence of the apnœa the excitability of the brain is much lowered, so that the same stimulus re-applied to the same spot produces a different change in the respiratory rhythm. Other conditions produce or add to the anæsthetic state, such as loss of blood, exposure of the brain, extravasated blood pressing on the brain, general exhaustion of the animal, or some disease present in the animal.   On the other hand, morphine or chloral in considerable doses completely obliterates some of the respiratory effects.

In the four species of animals, monkey, dog, cat, and rabbit, it is easy to form a judgment as to the state of anæsthesia—at least, after paying a little attention to the point.   The rapidity of the corneal reflex may act as a guide, or, better still, the excitability of the second division of the 5th nerve in the orbit, or of the dura mater.

GENERAL EXPERIMENTAL METHOD.

The essential preliminary to every experiment has been a regular respiratory rhythm, obtained by employing such a stage of anæsthesia as proved by experience to be best suited for the particular experiment proposed.   A tracing of the respiration and circulation before, during, and after the application of the electrical stimulus was taken in each experiment, and from the tracings alone have the conclusions furnished been drawn.

All the experiments have been repeated many times in each species of animal. Although, of course, only comparatively few tracings can be included in this paper, there is no point mentioned of which I have not many illustrations amongst the tracings in my possession.

The Age of the Animals.—In accordance with what has been known for many years (LEGALLOIS,[*] SEMON and HORSLEY,[†] &c.), the age of the animal is an important feature in the innervation of respiration.   I have, therefore, in all cases used adults only.

Record of the Respiratory Movements.— After anæsthetizing the animal a tube was inserted into the trachea, so as to exclude any mechanical obstruction to the respiration, or any complication of the result by a simultaneous excitation of the larynx, whilst, at the same time, by this means the anæsthetic can be more readily regulated. A PAUL BERT's transmission apparatus, connected with a MAREY's receiving tambour, was fixed by an inextensile band round the lower third of the thorax, where the circumference enlarges most during ordinary respiration.   The tracings, therefore, in the present research record the variations in the circumference of the lower third of the chest.   In all cases the tracings are to be read from left to right, the upstroke representing the enlargement in inspiration of the chest from the position of equilibrium, and the downstroke the return during expiration to the position of rest.

The following terms applied to the respiratory tracing will be made use of in order to distinguish between the very different degrees of respiratory movement observed.

---

* LEGALLOIS, 'Œuvres.'   Paris, 1830.   Vol. 1, 1.
† SEMON and HORSLEY, "The Central Motor Innervation of the Larynx." 'Philosophical Transactions,' 1890, B, p. 187.

Expiration.—The line drawn through the lowest points of the normal tracing.

Active Expiration.—Any line below expiration.

Inspiration.—A line drawn through the tops of the normal respiratory curve.

Over-inspiration.—Any line above inspiration.

Record of the Circulation.—The cannula was placed in the carotid of the dog, cat, and rabbit, and in the common iliac in the case of the monkey, and connected with a mercury manometer.

Although this was regularly done, I propose not to refer at length to the changes in the circulation obtained simultaneously with the respiratory ones, but to reserve my observations under this heading for a future communication. I have noted, however, on page 653, two well-marked effects which may be found.

The Stimulus.—This was obtained through a Du Bois-Reymond Inductorium from two Grenet's cells. The current was just sufficiently strong to be perceived by the tip of the tongue when the secondary coil was 12 centims. from covering the primary. The numbers 12, 10, 8, 6, 4, 2, 0 will be used as abbreviations to express the strength of the faradic stimulus when the secondary coil was 12, 10, 8, 6, 4, 2, 0 centims. distant from covering the primary.

The electrodes were of platinum, about 1 millim. apart.

An electromagnet in the primary circuit served to indicate the application and duration of the stimulus.

The rate of movement was marked by a metronome beating seconds, and recording by an electromagnet.

The recording points of respiration, circulation, and stimulus were kept vertically in line.

The recording apparatus used was the Kymographion of Hürthle.

The Course of an Experiment.—The animal being suitably anæsthetized, and the point to be stimulated exposed and gently dried, the electrodes were placed against the surface of the brain. A tracing was first taken with the electrodes in position before the stimulus, secondly, during the excitation, and thirdly, it was continued after the cessation of the stimulus until the respiration returned to its previous condition.

A current less than sufficient to produce any effect was first employed and then one of increased strength. The stimulus was applied for six seconds or more, and was often repeated at the same spot at short intervals.

Exposure of the Brain.—The skull was always removed so as to sufficiently expose the part to be excited, the bleeding being stopped by soft wax and amadou. The dura mater was raised as occasion required, and the brain kept warm and moist throughout the experiment by salt solution at blood heat, save that the spot was dried where the electrodes were to be applied.

The contents of the orbit were removed, or that of the eyeball only and the sclerotic drawn forwards.

In hemi-sections or in sections of both hemispheres, the plane of the surface was made perpendicular to that of the under-surface of the brain at the point of section, and also at right angles to the longitudinal axis.   In hemisections, particular care was taken to reach up to the middle line and not beyond.

A certain amount of hæmorrhage inevitably ensued in making such hemisections, but it was controlled by pressing the vessels against the base of the skull with sponges and amadou.

HISTORICAL RETROSPECT.

Alterations in respiration have been frequently noticed by various investigators during experiments upon the cerebral cortex and upon the basal ganglia, but the results obtained have not agreed together, or have been negative.

I am disposed to think that the variations in the results have been due to the use of non-anæsthetized animals in which the effects upon the respiration were complicated by other sensorimotor effects.   The negative results followed from not stimulating excitable spots, or because the animal had been too deeply anæsthetized by drugs or in other ways.

DANILEWSKY* used young cats and dogs slightly morphinized, and recorded the respiration by a tracheal cannula.   He obtained two kinds of results.   Cortical excitation in the region of the facial centre of HITZIG produced at first a slowing of the respiration with increased amplitude.   With a stronger stimulus a deep inspiration was followed by a longer expiration and a pause.   These phenomena only occurred sometimes, and often took place after the stimulus had ceased, even as much as 10 seconds after in one case.   A sufficiently strong excitation applied towards the base of the brain, especially when it affected the cerebral peduncles, caused marked acceleration of the respiration similar to the result of stimulating the dura mater and the peripheral nerves.   The arrest of respiration in these experiments may, from my observations, have possibly been due to the morphine, for slowing of respiration is observed to occur without any stimulus in dogs and cats well under the influence of morphine, and arrest may be produced in such animals by the excitation of any sensory part.

LÉPINE† laid stress upon the increased frequency of the respiratory movements.

BOCHEFONTAINE‡ produced rapid movements of inspiration, followed by spasms of varying intensity, and finally convulsions.   The experiments were made on non-anæsthetized animals.

RICHET§ employed dogs so deeply chloralized that the limbs did not react to cortical

---

* DANILEWSKY, "Experimentelle Beiträge zur Physiologie des Gehirnes." 'Archiv für Physiologie, Pflüger,' 1875, vol. 11, p. 128, see Table VI. and VII.

† LÉPINE—quoted by FRANÇOIS FRANCK, see below.

‡ BOCHEFONTAINE, 'Archives de Physiologie,' BROWN-SÉQUARD,' 1876, vol. 3, series 2, p. 168.

§ RICHET—quoted by FRANÇOIS FRANCK, see below.

excitations. He then found that stimulation of the sigmoid gyrus arrested respiration. The same result occurred not only at many other points of the cortex, but also was obtained from the sciatic nerve.

Both chloral and morphine, according to my own observation, tend to produce an irregular rhythm in which pauses occur at uncertain intervals. This change may sometimes be induced by any sensory stimulus.

MUNK's[*] observations in this respect were published in 1883, and reprinted in 1890. Other experimenters have not confirmed them. When the electrodes were placed upon the frontal lobe, some millims. in front of the frontal, i.e., the crucial sulcus, and somewhat lateral to its median end (just where a shallow longitudinal depression runs from before backwards), respiration was arrested with the secondary coil at 7 or 6. The arrest was in deepest inspiration, and the diaphragm in extreme tetanic contraction. Very often acceleration of the respiration preceded the inspiratory tetanus, during which, by deeper inspiration and less expirations, the thorax and diaphragm came to occupy gradually the maximal inspiratory position, and were finally fixed in that position. The abdominal muscles during this time remained relaxed. When the electrodes were placed on the under-surface of the frontal lobe, about the middle, there followed from stimulus 7 or 6 either a powerful tetanus of the abdominal muscles in a maximal expiratory phase, or the latter contracted with great frequency but only to a very small extent, and returned with short jerks to the position of rest. If the electrodes approached the bulbus or tractus olfactorius sneezing or cough was produced.

In the ape along the horizontal limb of the præcentral sulcus, and between it and the middle line, with the stimulus at 7 or 6, the thorax and diaphragm assumed an inspiratory position; further outwards than this fissure excitation produced a tetanus of the abdominal muscles.

As far as they go these observations are in the main confirmed by my own.

FRANÇOIS FRANCK[†] made a further examination of the subject and criticized the work of previous observers. He pointed out that the arrest of respiration produced by excitation of the cortex was frequently due to apnœa following accelerated respiration or was produced by the severe tetanic convulsions into which the animal was thrown. He further showed that under deep morphine or ether respiration frequently became irregular with pauses which were very likely to coincide with an electrical stimulus. He concluded that excitation of the motor area with a sufficiently strong and prolonged stimulus produces modifications in the respiratory movements, and that excitation of other parts of the cortex does not produce respiratory reactions except in the case of convulsions, i.e., when the effect of the stimulus spreads to the motor area or tracts. The respiratory rate quickens or slows without any correspondence between the spot stimulated and the effect produced. It is according to him rather

---

[*] MUNK, 'Ueber die Formationen der Grosshirnrinde,' Berlin, 1890, p. 164.

[†] FRANÇOIS FRANCK, 'Leçons sur les Fonctions motrices du Cerveau,' 1887.

the degree of intensity of the excitation or the degree of excitability of the cerebrum which appears to influence the frequency. The strongest stimulus produced slowing and exceptionally arrest, so that one cannot discuss centres for acceleration or for slowing.

A change in amplitude, augmentation, or diminution may be observed both with an increased and a decreased rate. The inspiratory state coincides with the greater amplitude and the expiratory state with the less. Special points for inspiration and expiration cannot be found upon the cortex, each of the points of the motor zone can produce the respiratory modifications indicated above.

UNVERRICHT* experimented upon dogs anæsthetised with morphine and recorded the respiration by means of an œsophageal cannula. With a faradic current less than sufficient to produce muscular contractions when applied to the ' motor ' area, he obtained in some of the experiments slowing of the respiration by excitation of an area on the third external convolution outside the orbicularis centre.

PREOBRASCHENSKY† experimented with dogs under morphine and with cats under chloroform and ether. One-third of the experiments were negative. In the dog he obtained arrest of respiration for two or three seconds immediately after opening the skull, using a very weak faradic current, but he never saw an active contraction of the respiratory muscles. In some cats he obtained arrest in inspiration, in others no effect. His two last experiments were negative. The area stimulated was the same as that investigated by UNVERRICHT, viz., on the anterior part of the third external convolution.

It is well known that centres influencing respiration have been inferred from excitation experiments upon the cerebrum, and have been called " Cerebral Respiratory Centres."

CHRISTIANI,‡ experimenting upon rabbits, concluded that an inspiratory centre was situated in the lateral wall of the third ventricle, in the optic thalamus, just in front of the corpora quadrigemina and the commencement of the aqueduct. The area was extremely limited, only 1 millim. square, from which an electrical, mechanical, or thermal stimulation evoked an arrest of the diaphragm in inspiration, or a remarkable acceleration of the respiratory rate, together with an increase in amplitude. The centre was symmetrical on each side. Excitation of the corpora quadrigemina, just below or to one side of the aqueduct of SYLVIUS, caused an arrest in expiration. These latter experiments were made after the removal of the hemispheres, corpora striata, and optic thalami by vertical transverse section through the commencement of the aqueduct. Thus the ' inspiratory ' centre had been removed before the ' expiratory ' one was excited.

* UNVERRICHT, "Ueber die Innervation der Athembewegungen," ' Kongress für innere Medicin,' 1888, vol. 7, p. 237.

† PREOBRASCHENSKY, "Ueber Athmungscentren in der Hirnrinde," ' Wiener klinische Wochenschrift,' 1890, p. 833.

‡ CHRISTIANI, ' Zur Physiologie des Gehirnes.' Berlin, 1885, chapter 1.

MARTIN and BOOKER* stimulated the corpora quadrigemina in rabbits by plunging the electrodes into the substance of the mesencephalon.   When the points were near the iter, acceleration was produced, which was changed into inspiratory arrest on applying a strong stimulus.   By plunging the electrodes deep into the pons they obtained arrest in expiration.   They experimented on two cats only and obtained the same inspiratory results as in the case of the rabbit.

The last-named observers, by plunging in the electrodes, appear to have stimulated the same points as myself on the surface of vertical transverse sections.   Their experimental results varied, perhaps because they did not suitably anæsthetize their animals.   CHRISTIANI, no doubt, stimulated in the line of the tract which I shall later on describe as causing acceleration and over-inspiratory tonus.   Again, the shock to the rabbit of making a vertical transverse section of the brain through the beginning of the aqueduct, renders the animal deeply anæsthetized.   CHRISTIANI, under these circumstances, obtained arrest of respiration probably in the line of the tract of fibres to which I shall refer as causing arrest when stimulated.

In view of the importance of this subject, I have repeated their experiments in the way the observers aforenamed performed them, and have obtained the same results.   But I fail to see why the presence of "respiratory centres" should have been assumed from these experiments, if the term "respiratory centre" be used in anything like the same sense as when applied to the representation of respiration in the floor of the 4th ventricle.   In fact, excitation experiments of this kind give results which are equally well or even better attributed to the results of stimulating fibres than of stimulating central mechanisms.

KNOLL,† experimenting more recently, does not say anything about the use of an anæsthetic.   He found that excitation of the optic thalamus and corpora quadrigemina caused quickening of the respiration.   A strong current applied to the under-surface of the anterior corpora quadrigemina caused spasm, especially of the eyes, tail, and flanks, the respiration becoming shallower or arrested.   He never obtained an expiratory arrest.   He concluded, from his experiments, that there was no respiratory centre in either the optic thalamus or in the corpora quadrigemina, but that the effect of excitation of these regions was to produce impulses in psychical or sensory tracts leading to true respiratory centres in the medulla and cord.

His experimental results and the conclusions drawn seem to me to follow from using non-anæsthetized animals.

* MARTIN and BOOKER, 'Journal of Physiology,' vol. 1, p. 370.

† KNOLL, 'Sitzungsberichte der Akademie der Wissenschaften,' Wien, 1885, vol. 92, sec. 3, p. 328.

DESCRIPTION OF PHOTOGRAPHS

On mature consideration, I think it best to give a description of the photographs first of all, in order that the position of the points stimulated may be the more easily understood.

On the photographs of brains and sections of brains have been indicated by marks the points where the several effects to be afterwards described in detail were obtained.

Rabbit's Brain (see Plate 57).

Photograph I*a*. A photograph of the under-surface of a rabbit's brain.

A ring has been placed upon the strip of cortex between the outer margin of the olfactory tract and the rhinal fissure just in front of the point where the sylvian artery crosses the fissure. The ring marks the point which will be referred to as the spot, on the application of the stimulus to which respiration can be arrested. It also shows the junction of the olfactory bulb and tract, marked by a square, the point from which over-inspiratory clonus was obtained.

Photograph I*b*. A photograph of the upper surface of a rabbit's brain, slightly enlarged.

The cross marks the point at which a branch from the anterior cerebral artery curves over from the mesial to the convex aspect. This is the place where the most marked acceleration can be obtained by the faradic stimulus.

Cat's Brain (see Plate 57).

Photograph II. A photograph of the antero-inferior and outer aspect.

A small ring is placed upon the apex of the olfactory lobe, *i.e.*, the antero-internal end of the triangular convolution formed by the outer border of the olfactory tract, the rhinal fissure and the furrow made by the sylvian artery. This is the place where arrest can be obtained.

An outer line delimits a semicircular area, at the border of which respiration could only be arrested by the strongest current used. The border line commences at the junction of the olfactory lobe and tract, crosses the supraorbital fissure some 3 to 4 millims. from its lower end and also the anterior composite lobe without extending as far as the anterior end of the coronary fissure. The line touches the anterior end of the anterior suprasylvian fissure, runs backwards parallel to and from 2 to 3 millims. above the rhinal fissure, nearly to the anterior end of the anterior ectosylvian fissure. The limit then turns downwards and joins the rhinal fissure about the point where the sylvian artery crosses. The outer part of the olfactory tract forms the lower border of the area.

The upper and inner end of the supraorbital fissure just beyond the limit of the photograph is the spot where acceleration is most marked.

The junction of the olfactory bulb and tract is also shown.

Dog's Brain* (see Plate 57).

Photograph III. A photograph of a dog's brain in the same position as that of the cat's.

A small ring has been drawn upon the olfactory lobe between the junction of the supraorbital with the rhinal fissures in front, and the sylvian artery behind, to indicate the centre of the area by the excitation of which arrest of respiration can be produced. An outer line marks the limit where respiration could be arrested only by the strongest current used. This line commences on the olfactory lobe about 4 millims. in front of the junction of the supraorbital and rhinal fissures, it crosses the rhinal fissure about the point where there is a small notch to be seen, and the orbital lobe, to the supraorbital fissure, some 4 millims. from its lower end. Behind this point it passes backwards and downwards to the point where the sylvian artery traverses the olfactory lobe and rhinal fissure. The outer part of the olfactory tract is the lower border.

The upper and inner end of the supraorbital fissure is just beyond the photograph ; it is the point where the faradic stimulus causes the most acceleration.

The junction of the olfactory bulb and tract is shown.

Monkey's Brain (see Plate 58).

Photograph IV. A photograph of the under-surface of a monkey's brain.

A small semicircle placed just external to the olfactory tract in front of the sylvian fissure indicates the centre of the area for respiratory arrest. A larger semicircle marks the limit within which arrest could be obtained with the strongest current used. Commencing at the outer border of the olfactory tract about the middle of its length, the limit passes directly outwards to the transverse portion of the H-like orbital fissure, thence along the edge of the shallow concavity which the orbital lobe forms, across the sylvian fissure, and turning inwards on the temporosphenoidal lobe reaches the outer side of the optic commissure, some 2 millims. or so behind the sylvian fissure.

PHOTOGRAPHS OF VERTICAL SECTIONS OF THE CEREBRUM.

Upon the photographs have been placed the following marks :—

A Small Ring.—This indicates the point on each half of the section, by the stimulation of which arrest was obtained.

A Cross.—This indicates the point on each half of the section, by the stimulation of which marked acceleration was caused.

---

* For the names applied to the cat's and dog's brain, see LANGLEY, "The Structure of the Dog's Brain," ' Journal of Physiology,' 1883–1884, vol. 4, p. 248.

A Square.—This indicates the point on each half of the section, by the stimulation of which over-inspiratory clonus was produced.

The Letter I.—This indicates the point on each half of the section, by the stimulation of which well-marked over-inspiratory tonus and tetanus was most easily obtained.

Vertical Sections of Cat's Cerebrum (see Plates 58 and 59).

Photographs V. to XV. A series of photographs, taken from Dr. HOWARD TOOTH's microscopical preparations of a cat's brain, upon which marks have been made as above.

During the course of the research, I made use of a series of photographs taken from Dr. HOWARD TOOTH's microscopical preparation of a monkey's brain, and also photographs made from sections of the brain of the rabbit and dog. The points were mapped in on these photographs, but, on account of the similarity shown to the results obtained in the cat, these photographs have not been reproduced.

THE DETAILS OF THE EFFECTS OBTAINED BY EXCITATION OF THE CORTEX CEREBRI AND OF THE SURFACES OF VERTICAL TRANSVERSE HEMISECTIONS OR SECTIONS MADE THROUGH THE CEREBRUM.

A. DIMINUTION OF ACTION.

(a.) *Slowing and Arrest of the Respiration obtained from the Cerebral Cortex.*

The Necessary State of Anæsthesia in Experiments upon the Cortex.—In order to obtain arrest of respiration by excitation of the area shown in the photographs, the animal must be narcotized so that no voluntary or reflex movements are evoked by the excitation. But a slight corneal reflex must still exist, and a strong stimulus still be able to disturb the respiration through the dura mater, fifth, or sciatic nerves. When these effects have been abolished, the animal is too deeply anæsthetized, and the cortex will be found no longer excitable to the strongest current used. But the excitability quickly returns if the animal be allowed to take a few breaths of air free from ether.

It is necessary to use ether. Morphine, except in very small doses, tends to abolish the excitability of the cortex, to render the respiratory rhythm irregular, and, moreover, any excessive dose is not so quickly got rid of as ether.

The cat may easily be maintained in the suitable stage of anæsthesia, especially when the experiment is confined to the arrest area, so that it is not awakened by excitation of the sensori-motor region, nor on excitation of the dura mater.

The rabbit presents somewhat greater difficulties, especially as there is only one very limited point on the cortex whence arrest of respiration can be evoked.

Dogs and monkeys are not so easily maintained in the same state as the cat, they

4 K 2

are apt to quickly pass into a too superficial or a too deep anæsthesia. But if an extremely small dose of morphine be injected subcutaneously half an hour before the commencement of the experiment, the animals can be kept in a more constant state of anæsthesia, resembling the condition produced in the cat by ether alone. The dose of morphine in this case must not exceed 0·01 gr. per 1 lb. (0·0014 grm. per 1 kilo.) live weight, injected subcutaneously, i.e., 1 minim per 1 lb. live weight of the 1 per cent. morphine hydrochlorate solution.

The rabbit however cannot be given morphine even in very small doses without depressing the excitability of the cortex so that no arrest can be obtained. In my first experiments I found the arrest area in the dog and monkey without morphine and I have not used any morphine at all in the cat.

The Necessary State of Anæsthesia in Experiments upon Vertical Hemisections or Sections.—The deepest ether anæsthesia in which the animal will breathe regularly is required when exciting the surface of sections in order to obtain arrest and to obliterate all the effects of the stimulus upon other fibres, e.g., those sensori-motor and commissural ones which increase the activity of the respiratory movements. Rabbits are the most difficult to get to breathe regularly in deep ether anæsthesia. Here also a minute dose of morphine, 0·01 grm. per 1 lb. (0·0014 grm. per 1 kilo.) live weight, may be tried in such animals as do not breathe freely in deep ether anæsthesia. The morphine aids in keeping the animal deeply anæsthetized with a smaller quantity of ether.

Slowing and Arrest of Respiration by Excitation of the Cortex Cerebri of the Cat. See Photograph II.

The Strength of the Faradic Stimulus.—At the centre of the arrest area the stimulus required to arrest respiration was generally 7, or 6, sometimes 8. Weaker currents were not quite strong enough to stop respiration but caused slowing. Currents weaker still had no effect. The further away the point stimulated was from the centre of the area, the stronger the stimulus required until the limit was reached where 0 was necessary to check the breathing. Beyond the limit no arrest occurred although slowing of the rhythm might be observed.

The form of the Arrest.—This was generally in full inspiration, or at some point midway between inspiration and expiration. At the centre of the area arrest occurred sometimes in active expiration or in expiration.

The arrest, especially when the stimulus was sufficiently strong and applied near the centre of the area, could be maintained for 6-10 seconds, and after the cessation of the stimulus respiration commenced with a slow rhythm and only gradually regained its former rate. The experiment could be repeated again and again either at the same place, or at various points of the area.

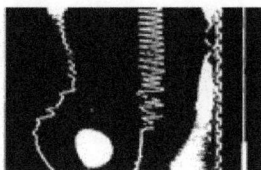

Tracing I.

Description of Tracings showing Slowing and Arrest in the Cat.

Tracing 1, p. 621. The point of the cortex excited was the central point of the area. See Plate 57, Photograph II.

The first line is the manometer tracing, the second that of the respiration, the third marks the rate of the travelling surface, the space between each vertical stroke being traversed in a second, the fourth the incidence, duration, and amount of the stimulus. With 6 the respiration was slowed or incompletely arrested, with 5 arrest took place in full inspiration, and this was repeated four times at short intervals.

Arrest in expiration could not be obtained so commonly in the cat as in the dog. The arrest in active expiration appeared to be limited to the centre of the area and was evoked by a current which did not influence any part beyond the exact spot mentioned. Moreover for this result it is essential that the effects of shock should be avoided as far as possible.

I will now give a table formed from tracings taken from most of the points of the area within which arrest of respiration can be obtained. All the tracings were taken from the same cat. Other excitation experiments made on the cortex of the same animal beyond the " arrest " area were negative. The arrest took place in full inspiration throughout, near the olfactory bulb the arrest was broken by the over-inspiratory clonus to be described later on. I also add a diagram made by enlarging Photograph II on which the numbers indicate the points stimulated.

Diagram 1.

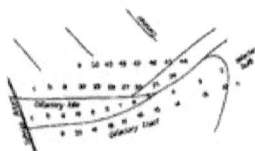

It is not maintained that there is a sharply defined limit beyond which a strong stimulus has no effect, but that the area described is the most extensive met with in favourable cases, the more deeply anæsthetized or the more exhausted the animal the more does the area contract.

TABLE.—See Diagram I. for points.

Amount of Faradic current.	Points resulting in	
	Arrest.	Slowing.
8	8 incomplete 9 " 26 27	7 10 (a) 10 (b) 25
7	6 7 8 9	
6	2 3 10 10 (a) 10 (b) 11 25 28 29 30 30 (a) 49	16 30 (b) 47 48
4	4 5 10 (a) 12 15 16 17 18 30 (b) 48	14 19 46
2	14 46	10 (c) 13 30 (c)
0	10 (c) incomplete 13 24 incomplete 30 (c) incomplete 45 50	20 44 50 (a)

Points.	Amount of current.	Result.
2	6	Arrest, "clonus."
3	6	"Clonus," then arrest.
4	4	"Clonus," then arrest.
5	4	Arrest.
6	7	Arrest.
7	8	Slowing.
7	7	Arrest.
8	8	Arrest, incomplete.
8	7	Arrest.
9	8	Arrest, incomplete.
9	7	Arrest.
10	7	Slowing.
10	6	Arrest.
10 (a)	8	Slowing.
10 (a)	6	Arrest, incomplete.
10 (a)	4	Arrest.
10 (b)	8	Slowing.
10 (b)	6	Arrest.
10 (c)	2	Slowing.
10 (c)	0	Arrest, incomplete.
11	6	Arrest.
11	4	"Clonus."
12	4	Arrest.
13	2	Slowing.
13	0	Arrest.
14	4	Slowing.
14	2	Arrest.
15	4	Arrest.
16	6	Slowing.
16	4	Arrest.
17	4	Arrest.
18	4	Arrest.
19	4	Slowing.
20	0	Slowing.
24	0	Arrest, incomplete.
25	8	Slowing.
25	6	Arrest.
26	8	Arrest.
27	8	Arrest.
28	6	Arrest.
29	6	Arrest.
30	6	Arrest.
30 (a)	6	Arrest.
30 (b)	6	Slowing.
30 (b)	4	Arrest.
30 (c)	2	Slowing.
30 (c)	0	Arrest, incomplete.
44	0	Slowing.
45	0	Arrest.
46	4	Slowing.
46	2	Arrest.
47	6	Slowing.
47	4	Arrest.
48	6	Slowing.
48	4	Arrest.
49	6	Arrest.
50	0	Arrest.
50 (a)	0	Slowing.

Slowing and Arrest of Respiration by Excitation of the Cortex Cerebri of the Dog.
See Photograph III.

The Necessary Stage of Anæsthesia.   See p. 619.

When the animal was strong a little chloroform was mixed with the first ether used, but not afterwards.   In any case ether has to be given in a more concentrated form to dogs than to other animals.   The employment of the small doses of morphine, mentioned on p. 620, is not absolutely necessary, and I did not use it until I had found the area, but it is certainly convenient for the prevention of sudden variations in the anæsthesia.

The Strength of the Stimulus.—At the centre of the area arrest was generally best obtained in the dog with 6, but in some instances 7 proved sufficient, others required 5.   Along the outer limit 0 was necessary, and there were intermediate stages where arrest took place with 4 and 2.   Just outside the border slowing could be produced with 0.

The Form of Arrest.—This was generally in active expiration, particularly at the centre of the area.   Less often the arrest was in expiration.   In the stage of anæsthesia employed it was only sometimes that arrest in any degree of inspiration occurred, but the tendency was more marked the further the distance from the centre of the area.

Tracing II., pp. 626 and 627.   For the three points excited see Photograph IIIa, b, c.

Arrest in active expiration was produced from three central points of the area excited in succession.   In each case 8 produced slight slowing, but 6 was required to arrest.   Some irregularity of rhythm with an increase in amplitude took place after the cessation of the excitation before the respiration resumed its regular character.

Tracing III., p. 630.   For the point stimulated see Photograph III.

The experiment in this case was made at the anterior border of the area and arrest was only produced at 0.   The arrest did not last the whole time of the stimulus, but during the latter part the breathing started again slowly.

Slowing and Arrest of Respiration by Excitation of the Cortex Cerebri of the Rabbit.
See Photograph Ia.

The Necessary Stage of Anæsthesia.   See p. 619.

Ether alone was, as a rule, used in this animal in sufficient amount to prevent voluntary and reflex movements from occurring when the spot was excited, but no more.   Even minute doses of morphine reduced the excitability enough to prevent any arrest of respiration from being obtained.

The Amount of the Stimulus.—At only one spot could arrest of respiration be obtained, and that with 8 or 6.   Stronger currents applied around this spot did not cause arrest.

Tracing II. (a) 1.

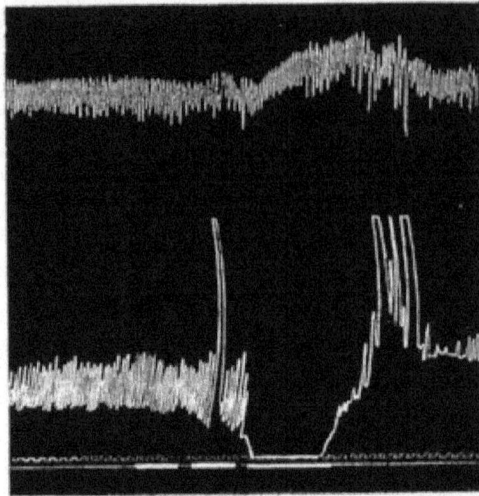

10   8   6

Tracing II. (a) 2.

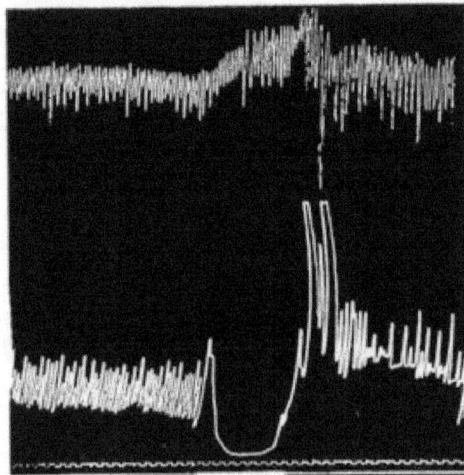

Tracing II. (b).

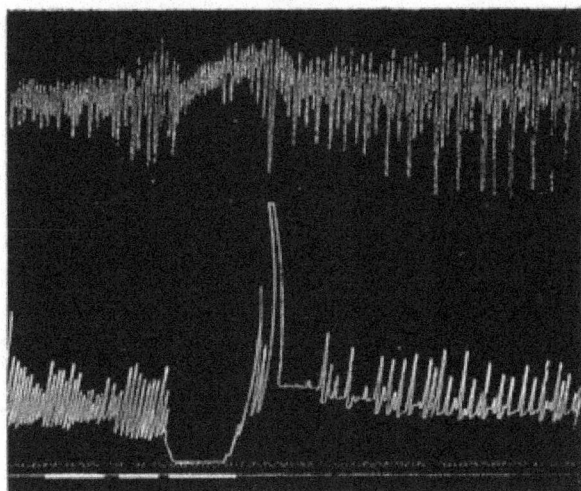

10   8   6

Tracing II. (c).

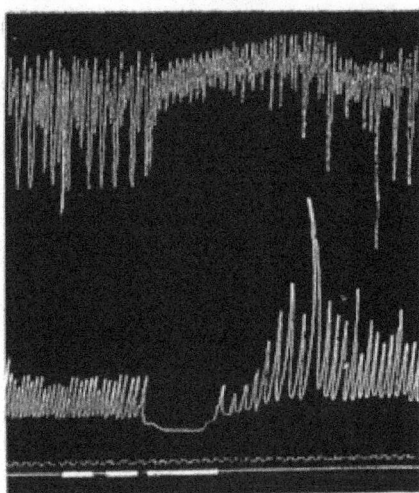

10   8   6

4 L 2

The Form of the Arrest.—This was nearly always in full inspiration, but as in the cat arrest in expiration occasionally happened.

Tracing IV., p. 630. The point excited is marked on Photograph I*a*.

The respiratory rate was slowed with 10, and arrested in full inspiration at 8.

Slowing and Arrest of Respiration by Excitation of the Cortex Cerebri in the Monkey (Macacus Rhesus).

The Suitable State of Anæsthesia. See p. 619.

The Exposure of the Orbital Surface of the Frontal Lobe.—The animal being placed on its back in a semi-recumbent position with the head fully extended, the orbital surface of the frontal lobe appeared facing forwards and upwards when the frontal bone including the orbital plate was removed.

The Blood-pressure Tracing.—The carotids were not interfered with in order to avoid complications from the effects illustrated by the experiments contained in a former paper.[*] Therefore the cannula in the monkey was invariably placed in the common iliac artery outside the peritoneum.

The Arrest Area.—This has been marked out in Photograph IV. I do not assert that the effect cannot be obtained at points internal to the chord of the arc drawn. But there are two difficulties in trying to decide this point, firstly, the liability of wounding the internal carotid artery and its branches, and secondly, the impossibility of preventing the spread of the excitation to the meninges. I prefer, therefore, not to absolutely delimit the inner border of this region.

The Degree of the Stimulus.—A current of 7 or 6 arrests respiration when applied just external to the junction of the olfactory tract with the uncinate convolution. But exciting points 1 millim. apart along any radius of the semicircle indicated in the photograph, arrest of respiration occurred with 4, 2, or 0, according to the distance from the centre. Just beyond the border 0 caused slowing, but not arrest.

The Character of the Arrest.—The arrest in the monkey was nearly always in expiration, but only rarely was any active expiration seen. Beyond the disappearance of the respiratory curves no change was noted in the circulation from excitation of the cortex only.

Tracing V., p. 631. The points excited are in connection with Photograph IV.

(*a*) Respiration which was not arrested by 8 was stopped by 7. This was at the centre of the area.

(*b*) 0 was required in this experiment, the point excited being at the anterior border of the area.

(*c*) A little within the limit 2 was necessary.

(*d*) Just outside the border there was no arrest with 0.

---

[*] SPENCER and HORSLEY, "The Control of Hæmorrhage from the Middle Cerebral Artery," 'British Medical Journal,' 1889, vol. 1, p. 457.

It is worthy of note that Tracing V., *a*, *b*, *c*, *d* were from the same monkey, some other similar experiments intervening.

The Occurrence of the Arrest.

It has, therefore, been shown that the respiration can be slowed and arrested by excitation of a certain spot and a limited area around it. This spot is situated in all the animals examined to the outer side of the olfactory tract just in front of the junction of the tract with the uncinate. And this arrest can be constantly obtained and the experiment repeated again and again under certain conditions.

1. The animal must be in such a stage of anæsthesia, that whilst the other effects upon respiration hereafter to be described are excluded, the excitability of the cortex for the arrest effect shall not be lost.

2. The animal must be in a normally active stage, and the brain to be excited must not have been injured in any way.

3. A sufficiently strong faradic stimulus is required.

Conversely the Causes which prevent the Occurrence of the Arrest.

1. (*a*) *Insufficient Anæsthesia.*—Over-inspiration and over-inspiratory tonus with slowing or acceleration of rhythm are excited when the narcosis is not of the right degree and arrest is prevented. In many cases the tracing thus obtained strongly suggested the struggle between the conflicting effects, *e.g.*, excitation sometimes produced a short arrest, which then became arrest in over-inspiration—and near the olfactory tract a clonic effect occurred instead of arrest. If the animal were allowed to awake sufficiently there might ensue reflex arrest for a second or two on exciting with a weak current, but a repetition of the excitation under these circumstances caused a different response.

1. (*b*.) *Too Deep Anæsthesia.*—With ether sufficient to abolish the corneal reflex, the cortex of this region was rendered inexcitable. At the centre of the area 0 might produce slowing but no arrest, even this was lost in deeper anæsthesia. Morphine, when given in larger doses than those mentioned, likewise abolished the excitability, the respiratory rhythm tended to become irregular, showing the phenomena of periodic respiration, of spontaneous pauses in expiration, and of convulsions. This was especially the case in the dog, but was also observed in the rabbit and monkey. When the animal was subjected to sufficient morphine for spontaneous arrests to occur, arrest of respiration, as others have noted, could easily be obtained by faradizing any sensory surface, including the convex surface of the brain, but was equally well got from a nerve such as the sciatic.

2. The cortex of animals not in a normal state of health before the experiment was more easily inhibited by the ether, and also the cortex was less excitable, *e.g.*, slowing, but no arrest, might be obtained. In other words, the depression due to imperfect health was added to the ether effect, and in some cases was sufficient

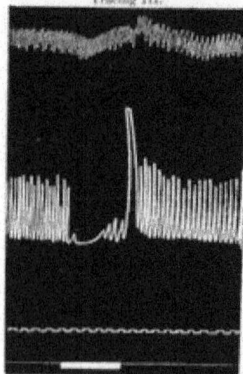

Tracing III.

0

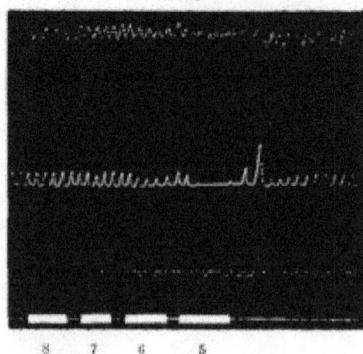

Tracing VI.

8   7   6   5

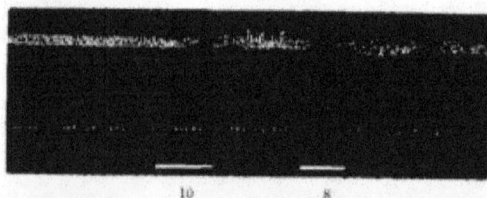

Tracing IV

10   8

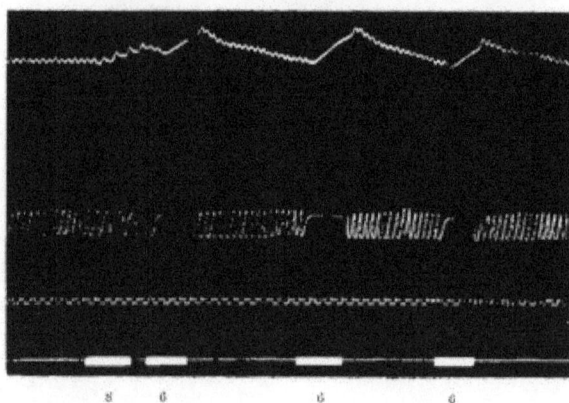

Tracing VII.

8   6   6   6

Tracing V. (b).   Tracing V. (c).   Tracing V. (d).

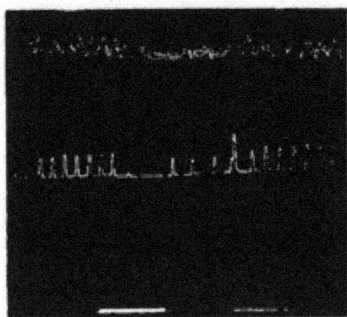 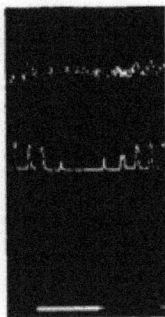 

0   2   0

of itself to abolish the excitability. Such instances occurred in monkeys with diarrhœa, in dogs with tape-worms, in rabbits with parasites in the liver, and in pregnant cats. The latter condition was not sufficiently advanced in any case to be noted before the experiment, and there was, of course, no mechanical obstruction to the respiratory movements. Some of the pregnancies noted were very early.

Exhaustion of the animal from exposure of the brain, from loss of blood, from repetition of the experiment, tended to diminish the excitability, and this ensued all the more rapidly the deeper the primary anæsthesia, whether from ether or from the prior causes above noted.

3. No arrest could be obtained except when the secondary coil was at least 8 or 7 centims. from the primary, although slowing of the rhythm might occur with a slightly weaker stimulus.

(b.) On the connection with the Medulla Oblongata of the area of the Cortex Cerebri for Slowing and Arrest of Respiration.

I will now proceed to describe the results of the excitation method I have employed to trace the fibres connecting the cortex with the central mechanism in the medulla oblongata.

On this point I need not describe in full detail the results obtained in each of the species of animals examined. Since the tract, connecting the area on the cortex with the medulla, runs through the ventral portion of the brain, i.e., through a part in which the difference between the four species is very slightly marked, and also because the results were practically identical. But although I have combined the description and inserted in this paper only a few tracings, yet I have, in my experiments, assumed nothing. I have stimulated every hemisection in each species of animal many times, and have tracings sufficient to afford many illustrations of every point.

Method of making Hemisections or Sections of the Brain. See p. 613.

The necessary state of Anæsthesia. See p. 619.

The deepest ether anæsthesia consistent with regular respiration was found to be the best. Of course great watchfulness was required to prevent an overdose. In some cases, when once arrested, the respiration did not begin until after some artificial respiration had been employed, but usually the respiration started again spontaneously. Only in some rabbits did the respiration tend to stop before they were sufficiently anæsthetized. I used, in these exceptional cases, the small dose of morphine already noted, viz., 0·01 grain per 1 lb. live weight, and so obtained the necessary anæsthesia without an excess of ether. As already mentioned, the dog required the ether vapour in a more concentrated form than other animals.

Tracing VIII. (a).

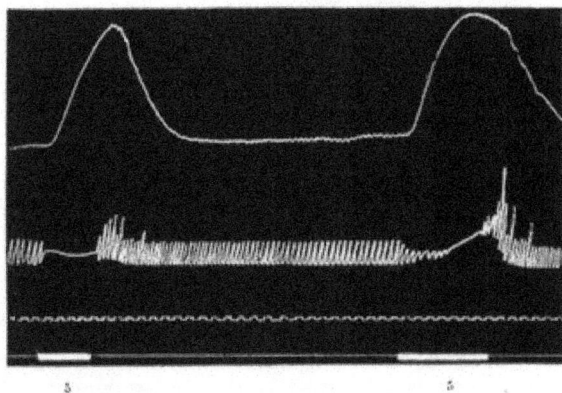

Tracing VIII. (b).

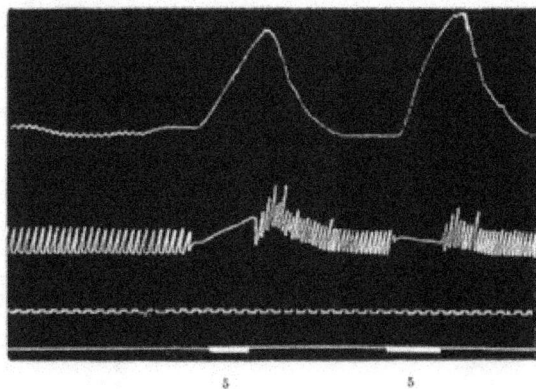

Mode of Experiment.—A vertical transverse (frontal) hemisection (sometimes a section) having been made in a deeply narcotized animal, and the point to be excited located and dried, the electrodes were held in position whilst a tracing of the circulation and respiration were being taken. The current was then applied for 6 seconds or more ; generally current 6, sometimes 7 or 5, was required to produce arrest. The record was continued until the recovery of the rhythm to the state existing before the experiment. When the electrodes were placed on the right spot and arrest obtained, the electrodes being held *in situ*, the experiment could be repeated again and again for six times or more with the same result.

The Form of the Arrest.—Arrest generally took place in full inspiration, less often in a position midway between inspiration and expiration. It was only in the deepest stages of anæsthesia that arrest took place in expiration, especially in monkeys. It was difficult to keep the other animals in this sufficiently deep stage whilst breathing regularly, or, if arrested by excitation, respiration required artificial aid to start again.

The Strand of Fibres which, when excited, produced Arrest.—If the series of photographs taken from Dr. TOOTH's sections of the brain of the cat be examined, and the fibres enclosed in the ring noted (see Plates 58 and 59), it will then be possible to trace the following description :—

From the centre of the cortical "arrest" area, in each animal fibres may be followed back, and these are observed to be gathered together at the lower and anterior end of the lenticular nucleus, and apparently form part of the bundle known in human anatomy as the olfactory limb of the anterior commissure. This bundle passes backwards, upwards, and inwards, to the inner side of the anterior cornu of the lateral ventricle. These fibres tend to the middle line, at which point they form the more anterior fibres of the anterior commissure. Evidence that these fibres actually decussate with those of the opposite side will be referred to directly. Behind the anterior commissure, the tract passes downwards and outwards from the middle line, close to the infundibulum, above the optic commissure, and then above the inner end of the optic tract. Behind this, it runs above and just internal to the crusta. At the beginning of the aqueduct, the strand runs backwards in the tegmentum, just external to the fibres of the 3rd nerve as they pass towards the ventral surface. At the level of exit of the 3rd nerve the fibres lie vertically above the point of exit of the nerve, being an equal distance below and outside of the aqueduct in the structure known as the red nucleus. The strands on either side are here parallel ; as far as the exit of the 3rd nerve the bundle of fibres can be traced microscopically, but, beyond, they become lost amongst the other fibres of the tegmentum. But evidence can be obtained experimentally of their parallel course as far back as the upper border of the pons. Beyond this I have not traced them. The animals used in the present research were adults, and in them, when a hemisection was made involving the pons fibres, the respiration either became irregular or very slow. It is possible that younger individuals would be more tolerant of disturbance. Two variations in the method of experiment went to show that the strand on each side really decussates at the anterior commissure. Thus (1) if the brain be hemisected frontally and removed as far back on one side as the origin of the 3rd nerve, the other half of the brain is thus exposed on its mesial aspect, so that either the cortical mesial surface or the cut surface can be excited. There is one point, and one point only, on the mesial aspect where arrest can be produced by excitation, and that is the cut surface of the anterior commissure, and immediately behind and below this structure. The fibres which have come from the cortex of the side removed, are in this way excited immediately after decussation. In the monkey the decussation of the so-called olfactory limb of the anterior commissure can be easily perceived to take place in front, and can be clearly distinguished from the rest of the anterior commissure which joins the temporo-sphenoidal lobes. Again, if after hemisection and extirpation at the level of the 3rd nerve the cortex on the opposite side be excited, no arrest can be obtained, nor can any arrest be obtained from hemisections made in this remaining half in front of the anterior commissure. By using very strong currents some slight slowing may occasionally be observed.

Description of Tracings to Show the Result of Excitation of Cerebral Hemisections in Producing Slowing and Arrest of Respiration.

Tracing VI., p. 630. Monkey. From a hemisection immediately behind the cortical arrest area. See Photograph V.   Coil 6 caused slowing with diminished amplitude ; 5, arrest in expiration.

In the dog, arrest in active expiration and in expiration was obtained with a tendency towards an inspiratory phase when a stronger stimulus was used.

Tracing VII., p. 630. Cat. From a hemisection at level of Photograph VII. where the "olfactory limb of the anterior commissure" first appears as a distinct bundle. Arrest occurred in inspiration.

Tracing VIII., p. 633. Rabbit. From a transverse section through the anterior commissure, i.e., the electrodes were applied to the anterior commissure in the middle line. See Photograph IX. The respiration was arrested four times in succession with 5. There was each time a great rise of blood pressure. It is to be noted that the rabbit could not, without risk, be always so deeply anæsthetized as the other animals. On the contrary, in the monkey this is easily accomplished, and then it will be seen that no rise of blood pressure occurred. See Tracing IX.

Tracing IX., p. 636. Monkey. From the mesial surface of the left hemisphere after the removal of the right half behind the level of exit of the 3rd nerve. The electrodes were applied just behind and below the cut surface of the anterior commissure. See Photographs IX. and X. Arrest was obtained in expiration twice with 6 and six times in succession with 5.

4 M 2

Tracing IX. (6).

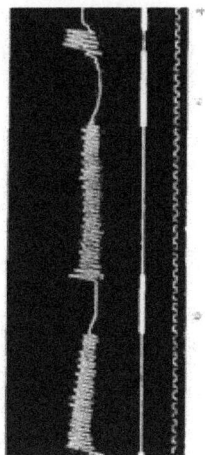

Tracing X. (b).

Tracings VIII., IX., and XI. illustrate an important point, viz., that the results described can be obtained many times in succession by keeping the electrodes to the same spot, and making use of a stimulus sufficiently strong, yet not so great as to quickly induce exhaustion, provided always that the animal be kept in the same state of anæsthesia.

In the case of the experiments illustrated by the next tracing, a variation in procedure was adopted, in order to show the influence of the etherisation. The animal was first allowed to come out a little from the state of deep anæsthesia, and then excitation was repeated whilst ether vapour was being administered, so that the animal was in a more deeply anæsthetized state with each experiment than it had been in the previous one.

Tracing X., p. 637. From a hemisection in a dog at the level where the optic tract is cut longitudinally in its course to the occipital cortex. See Photograph XI. The animal during the time that the record was being taken was breathing ether and becoming more deeply anæsthetized. Arrest was produced eleven times in succession at the same spot in an animal which was at the commencement in a superficial stage of anæsthesia, but, by breathing ether, gradually passed into a deep stage. In the first experiment the arrest was complicated by the over-inspiratory tonus, which will be referred to later on. In the second experiment the over-inspiratory tonus tending to complicate the arrest was again present, but weaker. The tracing also illustrates a commonly observed point, that with the deepening of the anæsthesia some increase in the strength of the stimulus was required. Thus the first experiment required 8 and the last 4, but allowance must be also made for exhaustion due to the repetition of the experiment at short intervals.

Tracing XI., p. 639. From a monkey at the level of exit of the third nerve, i.e., by exciting the centre of the red nucleus. Arrest was obtained at the point indicated six times in full inspiration. See Photograph XIV.

Tracing XII., p. 642. Rabbit. From a hemisection behind the level of exit of the third nerve and immediately above the pons. See Photograph XV. Arrest was obtained three times with 6 and four times with 5, each time in full inspiration.

Tracing XIII., p. 642. Dog. From the mesial surface of the left hemisphere behind and ventral to the anterior commissure after right hemisection and removal behind the level of exit of the third nerve. The arrest in quarter to half inspiration was maintained by 0 for 56 seconds. The tracing shows that the arrest of respiration could be maintained for long periods by using strong currents. Of course, exhaustion was readily produced in this manner.

Causes which Prevent the above Results from being obtained.—Unless the anæsthesia be sufficiently deep, over-action, such as acceleration and over-inspiratory tonus, is evoked and either prevents the arrest, or masks it. The electrodes must be placed exactly on the spot, and any pressure must be made in the long axis of the course of the tract, a very small deviation is sufficient to cause failure. When the

Tracing XL (b).

electrodes are applied exactly to the right place and held steadily there, the experiment may be repeated many times in succession. Exhaustion with slow or irregular rhythm does not abolish the excitability, but the arrest being produced, it tends to continue even after the cessation of the stimulus, and artificial respiration may be required to start the rhythm.

THE DETAILS OF THE EFFECTS OBTAINED BY EXCITATION OF THE CORTEX CEREBRI, AND OF THE SURFACE OF VERTICAL SECTIONS THROUGH THE CEREBRUM, IN THE CAT, DOG, AND RABBIT.

B. AN INCREASED ACTION.

(1.) *Acceleration of the Rhythm.*

General.—The excitation of any sensitive part in a non-anæsthetized animal tends to disturb the respiration in the direction of increased action, the rhythm is more or less accelerated, irregular, and there are muscular movements of the limbs and trunk. In a slightly anæsthetized animal a strong stimulus produces this effect over a wide extent of the cerebral surface. Under ether anæsthesia sufficient to abolish the excitability of the cortex so far that no movements of the limbs are evoked by the stimulus, acceleration of respiration is not caused by faradic excitation except by excitation of one spot on the cortex. Even if the faradic current be so strong as to call forth automatic movements, any acceleration of the rhythm which then may be obtained elsewhere is much less pronounced than the rapid rhythm provoked by exciting this particular spot.

On stimulating hemisections of the brain from before backwards this acceleration area on the cortex can be found to be connected with the medulla along a definite line, excitation of any point along this line producing marked acceleration.

So long, therefore, as the animal is sufficiently under the anæsthetic, acceleration of rhythm only occurs on the stimulus being applied at the one spot on the cortex and along the course of the tract through the cerebrum.

The Suitable Stage of Anæsthesia.—In all three species of animals on which the following observations were made, sufficient ether was given to abolish the excitation of movements of the limbs, and to suppress the over-inspiratory tendency which will be later on referred to. This applies both to the excitation of the cortex and to that of the surface of the sections of the hemisphere. A deep stage of anæsthesia abolished all cortical excitability for acceleration, as well as that of the sections.

One source of error had always to be carefully excluded, viz., the alteration in the respiratory centre due to apnœa produced by the acceleration. It was therefore found advisable to only allow the stimulus to act for about six seconds, otherwise the respiratory rhythm began to get slower. It was also necessary to allow full time for a return to the normal respiratory rate before repeating the experiment, otherwise the excitability was found to have been lowered by the apnœa.

The Area on the Cortex Cerebri where the most marked Acceleration was obtained on Excitation.

In the Dog and Cat.—The effect was best marked at the upper end of the supra-orbital sulcus. From a limited area around the acceleration could be obtained, but in a less marked degree. On increasing the anæsthesia this was the last place where any acceleration could be obtained with 0, and it was at this point that acceleration could be first obtained in an animal recovering from deep anæsthesia.

In the Rabbit.—If, when viewing the dorsal aspect of the rabbit's brain, the eye travels back from the olfactory bulb along the inner margin of the hemisphere, an artery is seen coming up between the edge of the hemisphere and the falx cerebri, and then turning over on the convex surface and running outwards in a line which suggests the position of the crucial sulcus in the cat and dog. When the pia mater is carefully stripped off, a groove remains. This vessel serves to indicate the place where marked acceleration can be obtained in this animal. When the electrodes are placed astride of the vessel near the margin marked acceleration is obtained. An area of 2 millims. in diameter overlapping the margin, i.e., extending a little way on to the mesial surface, is about the limit within which this phenomenon is markedly represented.

The Connection with the Medulla Oblongata of the Area on the Cortex which causes Acceleration when Excited.

The course which the fibres leading from the cortex cerebri take towards the medulla oblongata is indicated upon the accompanying photographs by a cross. It descends in the corona radiata to the lenticular nucleus where this structure blends with the lower and external part of the internal capsule as seen in frontal sections. The band of fibres then passes below the internal capsule as the latter changes into the crusta, and appears to slope towards the inner portion of the tegmentum. The strand on each side reaches the middle line in the same frontal plane as the exit of the 3rd nerve. The spot where acceleration is at this level obtained is thus in the inter-peduncular grey matter, rather nearer to the ventral margin than to the floor of the aqueduct.

The acceleration tract on either side appears to decussate in the plane of the exit of the 3rd nerve. I have already shown that hemisection and extirpation on one side behind the anterior commissure, but in front of the plane of exit of the 3rd nerve, removes the "arrest" effect; but marked acceleration can still be obtained from the spot on the cortex of the remaining hemisphere. On the other hand, removal of one hemisphere immediately in front of the pons does away with the special acceleration from the cortex of the remaining hemisphere. Some slight acceleration may still be obtained from the cortex and internal capsule if the anæs-

Tracing XIII.

Tracing XIV.

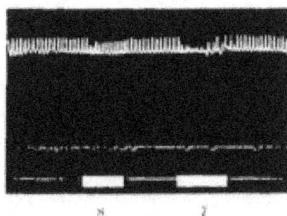

8   2

Tracing XVI.

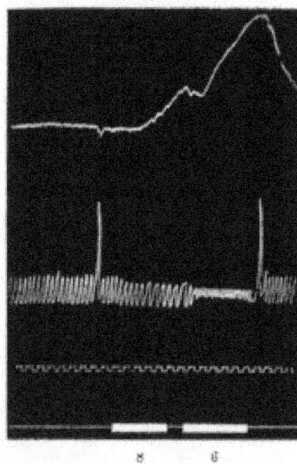

8   6

Tracing XV.

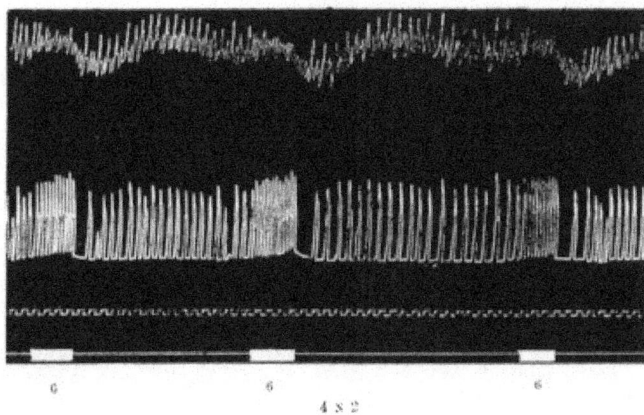

6   6   6

4 s 2

thesia become superficial, but this is the general effect mentioned on page 640, which is only very small, and is connected with the medulla through the descending sensori-motor tract, i.e., does not decussate above the pons. Finally, when one-half of the cerebrum has been removed no marked acceleration can be obtained from the mesial surface of the remaining half, except from a point immediately dorsal to and behind the exit of the 3rd nerve from the crus.

The Description of Tracings showing Acceleration.

The respiratory rate was counted by comparing a period of 6 seconds before and during the stimulus.

Tracing XIV., p. 643. Rabbit. The spot on the cortex cerebri of the right side was excited (see Photograph IB). Before any excitation the respiratory rate was 80 per minute. Stimulation with 8 caused the rate to become 100 per minute, i.e., an increase of 0·25. With 7 the rate became 120 per minute, i.e., an increase of 0·5.

Tracing XV., p. 643. Dog. Excitation of the upper end of the supraorbital fissure on the left side, after hemisection and extirpation of the right half behind the level of the anterior commissure. Excitation with 6 caused acceleration three times in succession, the first time from 50 to 110, an increase of 1·2, the second time from 40 to 120, an increase of 2, and the third time from 35 to 95, an increase of 1·7. The slow rate of respiration before the stimulus marks a deeper stage of anæsthesia than Tracing XXI. Further, also, an opposite hemisection having been made in this case, it is clear that the fibres connecting the cortex with the medulla cannot cross above the level of hemisection.

Identical effects were obtained from the cat.

Tracing XVI., p. 643. Cat. For point of excitation see Photograph VIII. With 8 no acceleration occurred, but with 6 the rate changed from 70 to 170, an increase of 1·4. There was a marked rise of blood-pressure; in other experiments with deeper anæsthesia no rise in blood-pressure occurred (see Tracing XVIII.), and the respiration being slower at the commencement, even greater acceleration followed. I have a tracing from a dog, taken from the same spot. 8 produced a fractional increase of 3·4, viz., from 25 to 110; the absolutely greater rate occurs in the slighter stages of anæsthesia, e.g., 170 in this Tracing.

Tracing XVII., p. 645. Rabbit. Right hemisection through the plane of the exit of the 3rd nerve, see Photograph XIV. With 8 the increase was from 60 to 120, i.e., an increase of 1 on the first experiment and an increase from 45 to 150 on the second experiment, i.e., an increase of 2·3. The second experiment was thus better than the first. It has often been found that an almost imperceptible movement of the electrodes may, on the one hand, improve the reaction and, on the other hand, impair it. This indicates the necessity of exciting the exact point if the maximal effect is to be obtained.

Tracing XVII.

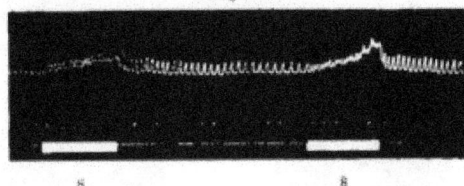

8   8

Tracing XVIII., p. 646.   Dog.   Right hemisection at the plane of exit of the 3rd
nerve.   See Photograph XIV.   The rate increased from 55 to 110, i.e., 1, with 10.
There was practically no variation in blood-pressure.   In this region where the electrodes
were so close to the crusta, the animal had to be well anæsthetized, although not so
deeply as required for the arrest of respiration, for if not, over-inspiratory tonus complicated the acceleration, and there was at the same time a marked rise in blood-pressure.

(2.) Over-inspiratory Clonus.   (" Snuffing" Movements.)

This movement produced by excitation was similar in the several species of
animals and was of the following character.   The animal made an over-inspiration,
and then several sharp over-inspirations were superimposed upon the primary one.
The over-inspiratory jerks were peculiar in following one another at regular intervals
in a rhythmic manner and in not ceasing exactly at the same time with the stimulus,
one, two, or three more of these over-inspirations taking place after the cessation of
the stimulus.

The result could be obtained from a definite area of the cortex, and along a tract
down to the medulla.

The Area on the Cortex for Over-inspiratory Clonus.

This was invariably and most easily obtained from the junction of the olfactory
bulb and tract.   By excitation of the portion of the frontal lobe, "procean lobe,"
lying immediately above and behind this spot the same effect could be obtained by
means of a stronger stimulus.   But the typical effect with the weakest excitation
was obtained at the junction referred to, both on the dorsal and lateral aspect.

The Apparent Course of the Fibres Connecting the Cortex with the Medulla by the Excitation of which Over-inspiratory Clonus is produced.

This was followed in the vertical sections commencing with those of the outer
limb of the olfactory tract, and continuing backwards past the furrow formed by the

Tracing XVIII.

Tracing XX.

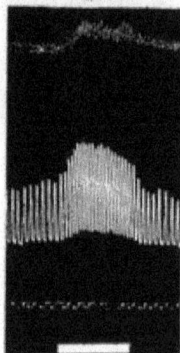

10

8

Tracing XIX.

Tracing XXI.

Tracing XXII.

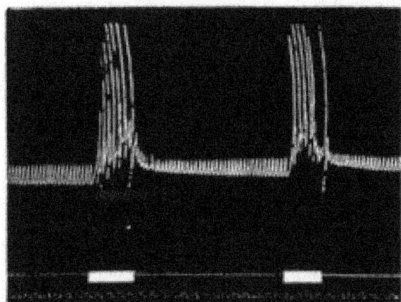

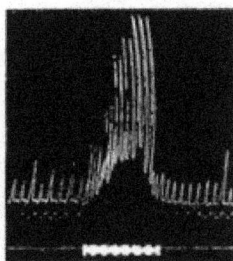

Tracing XXIII.

sylvian artery to the temporo-sphenoidal lobe at its mesial part, viz., the uncinate convolution. The effect obtained by excitation was traced along the uncinate convolution to the uncus. In the vertical section the over-inspiratory clonus is thus obtained from the region outside the optic tract as it passes backwards from the chiasma, whereas all the other respiratory effects are represented in this section internal to the tract. In the next section, behind where the optic tract has extended to the occipital lobe, the over-inspiratory clonus is to be obtained from the outer side of the peduncle, external to the crusta. Immediately behind the exit of the 3rd nerve transverse pontine fibres begin to separate the crusta from the ventral surface, and over-inspiratory clonus can here be obtained beneath the pyramidal tract. Behind the posterior perforated spot, at the naked-eye margin of the pons, the clonus can be obtained in the middle line at the ventral margin.

The course of the fibres which seem to subserve this reaction was easy to trace as far backwards as the section through the optic tract, see Photograph XII. Naturally, however, the connection between the uncinate gyrus and the region of the crusta, &c., was involved in greater difficulty. I repeated the experiments on this point whilst constantly referring, as I have always done throughout this research, to Dr. TOOTH's microscopical preparations of the cat's brain and to the photographs made from them. In the section from which Photograph XIV. was taken transverse or oblique fibres are already apparent beneath the crusta, although the 3rd nerve is not yet passed, and the posterior perforated spot still forms the ventral margin in the middle line.

The excitation of the under-surface of the anterior border of the pons was the one spot where the over-inspiratory clonus was obtained in the middle line. In agreement with this, the removal of one hemisphere through the plane of exit of the 3rd nerve did not abolish the effect obtained by excitation of the remaining hemisphere, although the arrest and acceleration effect disappeared.

The Degree of Stimulus.

8 and 6 may be taken as the usual amounts required. If the animal were in a very superficial stage of anæsthesia, less was needed. Currents of increasing strength up to 0 produced a similar but less marked result in the deeper stages of anæsthesia.

Causes which Prevented the Effect from being Obtained.

It disappeared in deep anæsthesia even with 0. As the animal became exhausted the effect was more and more incomplete.

Over-inspiratory clonus might be complicated with arrest or with over-inspiratory tonus. It was liable to be complicated with arrest when the cortex was excited, owing to spreading of the stimulus or the effect of the same from the junction of the olfactory bulb and tract to the adjacent arrest area. Over-inspiratory tonus tends to affect the result in the sections behind the optic tract, where the strand passes ventral to the centre of the crusta. This was avoided by keeping the electrodes close to the under-surface of the crus, by not employing any stronger current than was absolutely necessary, and by using sufficient ether to diminish the tendency to over-inspiratory tonus.

Description of Tracings showing Over-inspiratory Clonus.

Tracing XIX., p. 646. Cat. Junction of olfactory bulb and tract, right side, see Photograph II. Over-inspiratory clonus was produced by 6 and 4, at first, also with 2, but afterwards arrest occurred from the extension of the stimulus to the arrest area. With this tracing may be compared the table on p. 623, as showing the effect of exciting the points marked 2, 3, 4, and 11 on Diagram I.

Tracing XX., p. 646. Dog. Junction of left olfactory bulb and tract, see Photograph III. The right hemisphere had been excluded by section between the plane of exit of the 3rd nerve and the anterior end of the pons. 8 produced the effect.

Tracing XXI., p. 647. Rabbit. Junction of left olfactory bulb and tract, see Photograph I b. The right hemisphere had been excluded by section at the plane of the optic tract. The clonus was twice elicited by a current of 0. The animal was in a deeper stage of anæsthesia than in the preceding cases. In a stage deeper still the reaction is lost.

Tracing XXII., p. 647. Dog. Tip of left uncinate convolution, see Photograph XI. and XII., and just external to the optic tract, after section of the right side between the plane of exit of the 3rd nerve, and the anterior end of the pons. The clonus was produced with 6.

Tracing XXIII., p. 647. Cat. The middle line at the ventral margin of the anterior end of the pons, see Photograph XV., after a right hemisection and removal of the hemisphere at this level, the clonus was produced with 10 and 8.

(3.) Over-inspiratory Tonus.

This is the most widely generalized effect of any stimulus upon the respiration. In a non-anæsthetized or partly anæsthetized animal, all previous observers have shown that a stimulus, if strong enough, will influence the respiration in this direction when applied to any sensory surface.

The greater relative influence of anæsthesia upon the over-inspiratory tonus has allowed of the other effects being largely or entirely freed from its complicating influence.

No doubt greater degrees of over-inspiratory tonus might be most easily produced on a non-anæsthetized animal, but all my experiments have been with anæsthetized animals; and I may recall what I have mentioned before, viz., that there are other sources of anæsthesia than drugs, and in order to produce the same amount of over-inspiration in the same animal with the same stimulus applied to the same place, it is necessary that the animal should be in a definite stage of anæsthesia, whether caused by ether, shock, hæmorrhage, &c. I have examined the production of over-inspiratory tonus from three points of view, by excitation of the 5th nerve and meninges, of the sciatic nerve after the removal of the cerebrum, as well as the cortical effects.

Over-inspiratory Tonus from Excitation of the 5th Nerve and Meninges.

In animals under sufficient ether to abolish voluntary movements, I only obtained by faradic excitation of the 5th nerve and meninges over-inspiratory tonus, and never observed any arrest of the respiratory rhythm. Slight excitation of the terminal ends of the 5th nerve in the nose, by chloroform or tobacco smoke, have been described as instances of the causation of arrest by excitation of this nerve. The well-known fact that every child or animal forced to inspire an anæsthetic, "holds its breath," is an example of this, and it requires a distinct effort of the adult human will to resist the tendency. This arrest through the peripheral terminations of the 5th nerve disappears when there is sufficient anæsthesia to abolish voluntary movements. Arrest has been produced in a like fashion by exciting the trunk of the nerve or its branches in the dura mater in a non-anæsthetized animal with a very weak current But an animal may be made "to hold its breath" by gentle excitation of many other sensory surfaces, such as the cerebral cortex or the sciatic nerve. In these cases, however, the arrest usually ends in a general convulsion, and on the repetition of such an experiment the convulsion is just as likely to occur primarily instead of an arrest.

When the anæsthesia is sufficient to abolish voluntary movements, then excitation of the 5th nerve root, of the Gasserian ganglion, of the divisions of the 5th nerve, and of the dura mater, results only in over-inspiratory tonus, and not in absolute arrest of the rhythm. If excessive currents are employed, the rhythm may be arrested by spreading from the dura mater to the arrest area of the cortex either immediately by direct contact, or mediately through the cerebrospinal fluid, or blood clot. It may also spread to the roots of the vagus. The amount of current required to produce over-inspiratory tonus varies with the amount of anæsthesia. In deep ether anæsthesia there was not the slightest effect produced upon the respiration even with the strongest current used. When the current was strong relative to the amount of anæsthesia, the thorax assumed an extreme degree of expansion, and the lever of the MAREY's tambour no longer recorded anything but fine tetanic movements, but on direct observation the respiratory rhythm was seen to continue although small in amplitude and masked by the extreme expansion.

Description of Tracings from the 5th Nerve, showing Over-inspiratory Tonus.

Tracing XXIV., p. 651. Rabbit. Excitation of the 2nd division of the 5th nerve in the orbit. The result was over-inspiratory tonus, with a continuance of the rhythm.

Tracing XXIV.

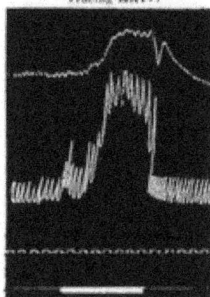

14

Tracing XXVII.

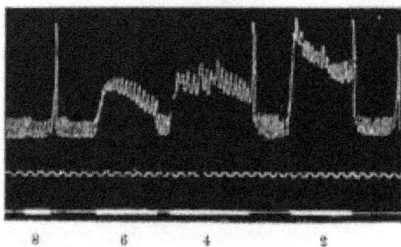

8   6   4   2

Tracing XXVI.

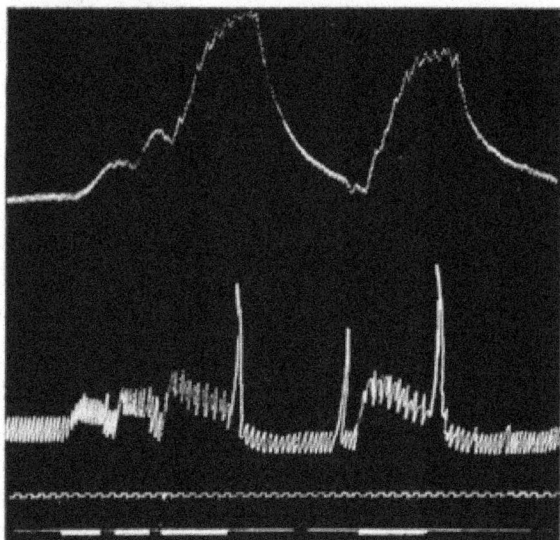

12   10   8   6

4 0 2

Tracing XXV., p. 647. Cat. Excitation of Gasserian ganglion. Marked over-inspiratory tonus occurred, and a diminution in amplitude from the expansion of the thorax which masked the respiratory movements.

In Tracings XXIV. and XXV. the same current was used; and other tracings from the dog and monkey, with the same strength of stimulus, show similar effects, only rather less marked in the case of the monkey. In comparing a number of experiments upon the 5th nerve in the various species, it is to be noted that when the anæsthetic was insufficient acceleration of rhythm took place, and continued after the cessation of the stimulus, until the animal was given more of the drug, when, on repetition of the experiment, the rate was not altered. Also, in exciting the 5th nerve, or dura mater, near the olfactory bulb, care had to be taken that the current did not spread through the olfactory nerves to the bulb, and set up "clonus."

In deeper stages of anæsthesia the same result occurs, but a stronger excitation is required.

Over-inspiratory Tonus after Removal of the Cerebrum.

In order to show that over-inspiratory tonus is a general effect of exciting any sensory nerve, and has no special connection with the cerebrum, I excited the sciatic nerve after the complete removal of the cerebrum by incision at the level of the tentorium cerebelli.

Tracing XXVI., p. 651. Cat. Excitation of the left sciatic nerve in the popliteal space, after the removal of the whole of the cerebrum immediately above the pons.

Over-inspiratory tonus occurred with 12, 10, 8, and 6.

The Area on the Cortex where the most marked Over-inspiratory Tonus is produced, and the apparent course of the Fibres connecting this Area with the Medulla.

The greatest effect is to be obtained in the centre of the "sensori-motor" area. In the sections the best effect is obtained from the centre of the descending motor tract in the internal capsule, and lower down, in the middle of the crusta. These fibres do not decussate early, for if the opposite hemisphere be removed just behind the plane of exit of the 3rd nerve, the effect may still be obtained by stimulating their course downwards in the vertical sections of the remaining half.

Whether there is one spot in the sensori-motor area more easily affected than any other, I cannot say, so readily is the result influenced by the anæsthesia.

Tracing XXVII., p. 651. Cat. Excitation of the "sensori-motor" area.

Great over-inspiratory tonus was produced with 6, 4, and 2.

The same effect followed excitation of this area when the opposite hemisphere had been completely removed down to the tentorium cerebelli.

Tracing XXVIII., p. 646. Dog. Excitation of the crusta just behind the internal capsule, see Photograph XIII. 8 produced over-inspiratory tonus, the rhythm continuing.

Comparing a number of tracings of over-inspiratory tonus, I find that, although best marked in the sensori-motor area of the cortex, and in the line of the descending motor tract, yet it is produced outside this and complicates the "arrest" or "acceleration" effect, unless removed by the anæsthesia.

Thus, on the "arrest" area, arrest in over-inspiratory tonus follows excitation when the anæsthesia is insufficient. Conversely it may be obtained from the "arrest" area, uncomplicated by any arrest after opposite hemisection behind the anterior commissure, and from the "acceleration" area uncomplicated after a similar hemisection behind the third nerve exit.

Tracing XXV.

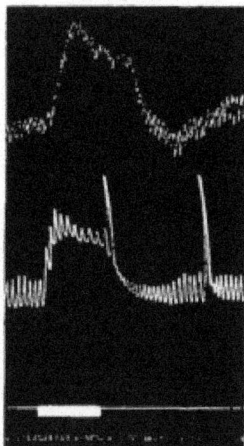

Tracing XXVIII.

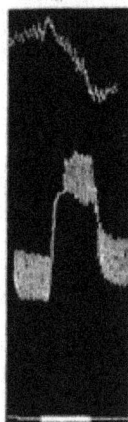

An Increased Action of the Respiration in the Monkey.

I have been forced to distinguish the monkey from the rabbit, cat, and dog, because in it an increased action of the respiration is so much less marked. There appears to be no fundamental difference, for I have obtained the same reactions, "acceleration," "over-inspiratory clonus," and "over-inspiratory tonus," at the points corresponding to those of the other animals, but only to a much smaller degree. The monkey reacts like the other three species do when very exhausted. I have not yet succeeded in artificially producing a greater excitability of the monkey's brain so as to obtain marked increased action of the respiration.   The result of such attempts has been to excite irregularity of the rhythm and general convulsions.   This lessened representation of the other phenomena, or the relatively greater sensitiveness to the effects of anæsthesia on the part of the monkey, allows slowing and arrest to be obtained all the more readily, and the localisation of the points of representation to be more easily recognised, since the effect is not so liable to be complicated by a simultaneous calling forth of increased action.

On the cortex over-inspiratory tonus and acceleration are to be obtained at the anterior end of the sulcus known as X (sulcus frontalis superior).

A Note of the Concurrent Effects upon the Circulation.

The stimulus directed to producing respiratory changes sometimes influenced the circulation, and notably in two ways :—

(1.) A marked rise of blood pressure occurred, especially when the sensori-motor area of the cortex was excited and the sections in the course of the descending motor tract.   But this happened only when the animal was not deeply anæsthetised. With deeper anæsthesia a very slight rise of blood-pressure took place, or no change at all.

(2.) In the region of the "arrest" area, especially in the neighbourhood of the sylvian artery, a fall of blood-pressure with a slower heart rate was met with.

But the exact conditions under which the circulatory effects can constantly be obtained has not yet been worked out.

CONCLUSION.   (See Diagram II.)

It has proved itself an easy matter to fail in obtaining an effect upon the respiration by excitation of the cerebrum, either from experimenting upon inexcitable parts, or owing to excessive anæsthesia however produced.

Again, in all animals, when too slightly anæsthetized, complex and variable results are liable to occur, such as combinations of slowing and over-inspiratory tonus, acceleration changing to slowing, arrest interrupted by over-inspirations.

But, by careful regulation of the anæsthetic state, the following definite results follow constantly from faradic excitations.   (See Diagram II.)

A. DIMINUTION OF ACTION.

Slowing and Arrest.

The cortical area, where this result was obtained, is situated just outside the olfactory tract in front of the point where the tract joins the temporo-sphenoidal lobe.

Followed back by vertical sections, the same result was obtained along the line of the strand of fibres known as the olfactory limb of the anterior commissure. After decussation in the latter structure, the tract continued backwards by the side of the infundibulum into the red nucleus below and external to the aqueduct in the plane of exit of the 3rd nerve.

Diagram II.

1. Cortical area for slowing and arrest.
2. Olfactory limb of anterior commissure.
3. Anterior commissure.
3(a). Fibres from the temporo-sphenoidal lobes.
4. Infundibulum.
5. Fibres passing backwards, causing slowing and arrest.
6. Red nucleus.
7. Exit of 3rd nerves.
8. Anterior fibres of pons.
9. Junction of olfactory bulb and tract.
10. Olfactory tract.
11. Uncinate convolution.
12. Uncus.
13. Connection of uncus and outer part of crus behind optic tract.
14. Optic tract and commencing optic radiation.
15. Tract passing obliquely beneath crusta.
16. The same, meeting the opposite one in the middle line at the anterior border of the pons.
17. Cortical area for acceleration of rhythm.
18. Tract external and ventral to internal capsule.
19. Tract passing below the junction of the internal capsule and crusta, above the optic tract, to the tegmentum.
20. Meeting of tract of either side in interpeduncular grey matter, at the level of and just behind the exit of the 3rd nerve.
21. Cortical areas for over-inspiratory tonus.
22. Tract passing through the motor portion of the internal capsule dorsal to the optic tract.
23. Tract running in the crusta.

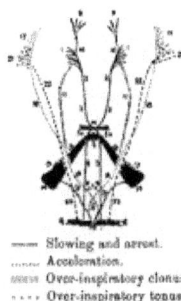

——— Slowing and arrest.
.... Acceleration.
▨▨▨ Over-inspiratory clonus.
· · · · Over-inspiratory tonus.

B. Increased Action.

(1) *Acceleration.*

Commencing especially from a point on the convex surface of the cortex within the sensori-motor area, the effect may be followed back through the lenticular nucleus where it borders on the outer and ventral portion of the internal capsule; the strand runs at first externally and then ventrally to the motor portion of the internal capsule, and so reaches the tegmentum. The lines from the two sides meet in the inter-peduncular grey matter at the level of and just behind the plane of the 3rd nerves.

(2) Over-inspiratory Clonus.

This effect was obtained from the junction of the olfactory bulb and tract, and on continuing to apply the stimulus backwards along the olfactory tract was traced into the uncinate convolution of the temporo-sphenoidal lobe. Followed to the uncus it passed behind the optic tract to the crus, and then pointing obliquely inwards ventrally to the crusta the effect on each side converged to the middle line at the upper border of the pons.

(3) Over-inspiratory Tonus.

The descending motor tract yielded this result, so did the 5th nerve and dura mater, as well as the sciatic nerve after complete removal of the cerebrum at the level of the tentorium cerebelli.

So far as the anatomical arrangement of the tracts above described is concerned, I may say that medullated fibres are to be seen in Dr. Tooth's preparations of the cat's and monkey's brain, running in the same course as that indicated by faradic excitation of living sections. An exception to this statement occurs in the case of the over-inspiratory clonus, in which I have not made out the connection between the uncus and the pontine fibres lying ventral to the crusta.

It is manifest that the respiratory alterations described above as capable of being induced by the faradic current are such as can be produced in the conscious state by volitional effort, and any explanation of the results described above must depend upon the acceptation or refusal of the general doctrine concerning the sensori-motor functions of the cortex, and that concerning the mechanism of the respiratory centre in the medulla.

I sum up therefore, the contents of this paper, when I say that whilst the effect upon respiration of exciting the cerebrum in a non-anæsthetized animal is probably a complex one, yet, by careful regulation of the anæsthetic state, four constant effects can be obtained upon respiration by stimulation of the cortex cerebri, and these can be traced down each in a course of its own from the cortex to the medulla oblongata.

DESCRIPTION OF PLATES.

PLATE 57.

Photograph IA.—Under-surface of a Rabbit's Brain, the small ring marks the point for slowing and arrest, the square that for inspiratory Tonus.

Photograph IB.—Dorsal surface of a Rabbit's Brain, the cross marks the point for acceleration.

Photograph II., with diagram.*—Ventral and lateral surface of a Cat's Brain, the inner ring marks the centre, the outer the boundary of the area for slowing and arrest.

Photograph III., with diagram*.—Ventral and lateral surface of a Dog's Brain, the inner ring marks the centre, the outer the boundary of the area for slowing and arrest.

PLATE 58.

Photograph IV., with diagram.*—Ventral surface of a Monkey's Brain, the inner ring marks the centre, the outer the boundary of the area for slowing and arrest

†Photographs V. to VIII.—Vertical transverse sections of Cat's Brain.

PLATE 59.

Photographs IX. to XV.—Vertical transverse sections of Cat's Brain.

* On the Diagrams appended to Photographs II., III., and IV., the Roman figures refer to the number of the tracings, and the dotted line from each number leads to the point of the brain, by excitation of which the tracing so numbered was obtained.

† The following marks have been placed on Photographs V. to XV., indicating the point on each half of the section by the stimulation of which was obtained:—

(1.) Slowing and arrest	. . . .	Marked by a small ring.
(2.) Acceleration	. . . . . .	,, ,, cross.
(3.) Over-inspiratory Clonus	. .	,, ,, square.
(4.) Over-inspiratory Tonus.	. .	,, ,, letter I.

Photograph I a
RABBITS BRAIN

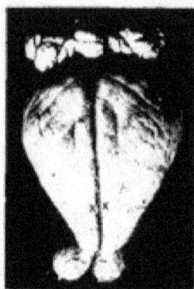

Photograph I b
RABBITS BRAIN

Photograph II
CATS BRAIN

Photograph III
DOGS BRAIN

DIAGRAM
to Photo II

DIAGRAM
to Photo III

Photograph IV
MONKEY'S BRAIN

DIAGRAM
to Photo IV

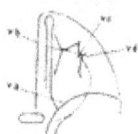

Photograph V

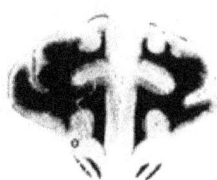

Photograph VI

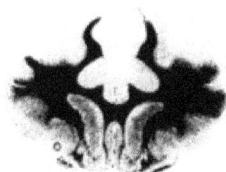

Photograph VII

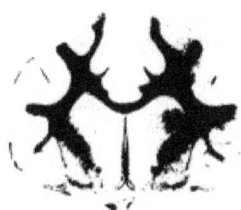

Photograph VIII

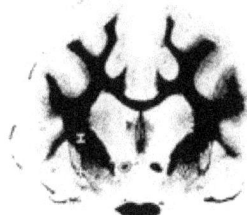

Photograph IX.

Photograph X.

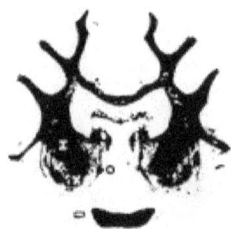

Photograph XI.

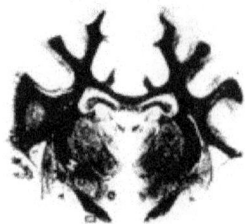

Photograph XII.

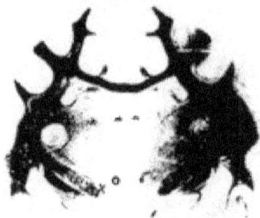

Photograph XIII.

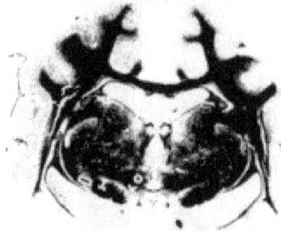

Photograph XIV.

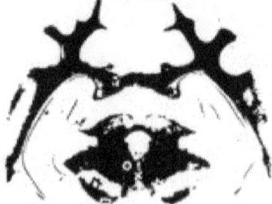

Photograph XV.

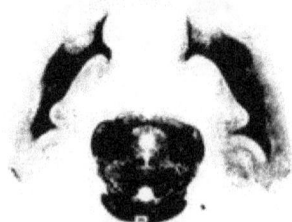

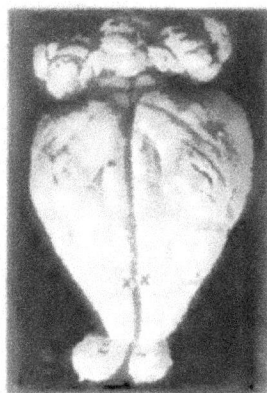

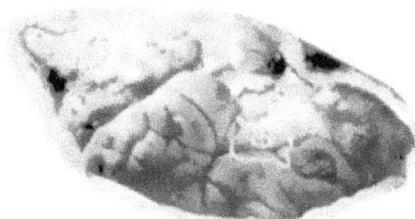

I.

PLATE 57.

Photograph I<small>A</small>.—Under-surface of a Rabbit's Brain, the small ring marks the point for slowing and arrest, the square that for inspiratory Tonus.

Photograph I<small>B</small>.—Dorsal surface of a Rabbit's Brain, the cross marks the point for acceleration.

Photograph II., with diagram.*—Ventral and lateral surface of a Cat's Brain, the inner ring marks the centre, the outer the boundary of the area for slowing and arrest.

Photograph III., with diagram*.—Ventral and lateral surface of a Dog's Brain, the inner ring marks the centre, the outer the boundary of the area for slowing and arrest.

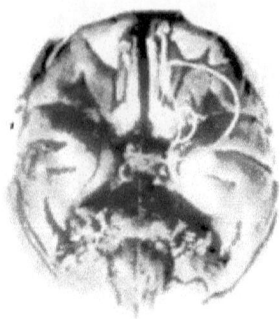

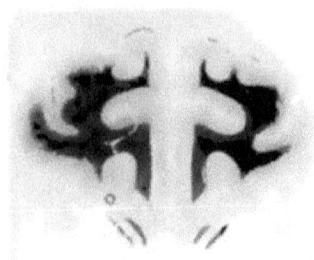

DIAGRAM

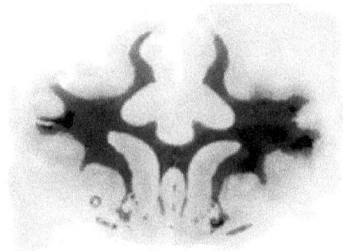

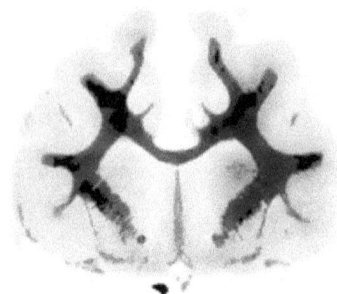

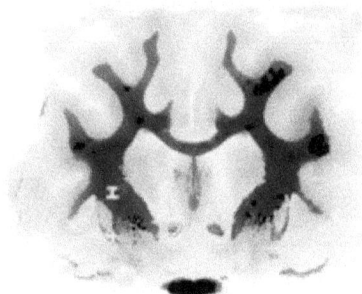

PLATE 58.

Photograph IV., with diagram.*—Ventral surface of a Monkey's Brain, the inner
   ring marks the centre, the outer the boundary of the area for slowing and
   arrest

†Photographs V. to VIII.—Vertical transverse sections of Cat's Brain.

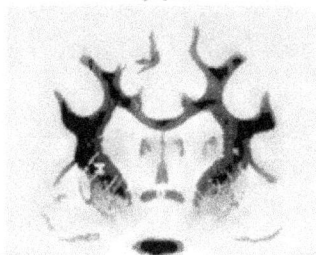

Photograph IX

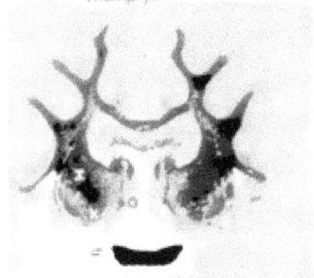

Photograph X

Photograph XI

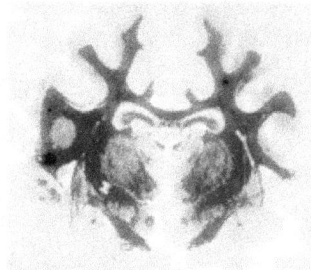

Photograph XII

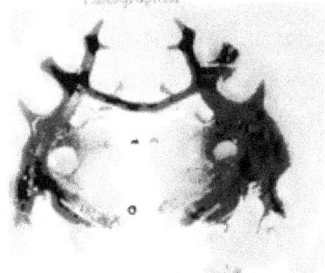

Photograph XIII

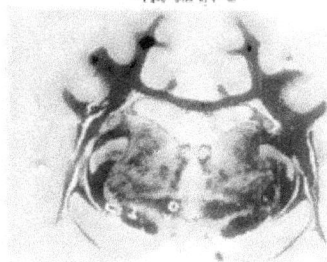

Photograph XIV

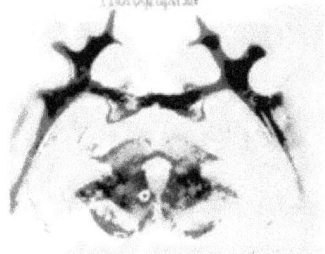

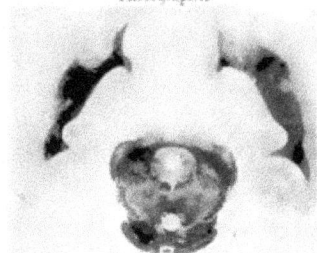

Photograph XV

PLATE 59.

Photographs IX. to XV.—Vertical transverse sections of Cat's Brain.